SQUEAKY TOYS

A COLLECTOR'S HAND BOOK & PRICE GUIDE

L.H. MACKENZIE

Schiffer Publishing Ltd
4880 Lower Valley Rd. Atglen, PA 19310 USA

DEDICATION

To my kids—Holly, Kerry, Scott, and Kurt—and all kids, young and old, that enjoy collecting toys from their early years.

Library of Congress Catalog Card Number: 98-85081

Designed by Bonnie M. Hensley
Typeset in Lithograph BT (bold)/Humanist 521 BT

ISBN: 0-7643-0622-7
Printed in China
1 2 3 4

Published by Schiffer Publishing Ltd.
4880 Lower Valley Road Atglen, PA 19310
Phone: (610) 593-1777; Fax: (610) 593-2002
e-mail: schifferbk@aol.com

In Europe, Schiffer books are distributed by Bushwood Books
6 Marksbury Avenue Kew Gardens
Surrey TW9 4JF England
Phone: 44 (0)181 392-8585; Fax: 44 (0)181 392-9876
e-mail: bushwd@aol.com

Please write for a free catalog.
This book may be purchased from the publisher.
Please include $3.95 for shipping. Please try your bookstore first.
We are interested in hearing from authors
with book ideas on related subjects.

ACKNOWLEDGMENTS

I would like to take this opportunity to thank the people who helped make this book possible. My thanks to Patty McWilliams, F. Doane, Joan Pappas, Joseah Lupton, and many more who shared their time and talents. Thank you all

INTRODUCTION

The love of antiques and what we collect comes to all of us in different ways. For me it started in my adolescence while attending an auction with my mother, who worked as a clerk for a prominent auctioneer in Keene, New Hampshire. My love of toys has developed in recent years due to the lack of toys as a small child. The ones we did have hold great memories, especially those from the late '30s and '40s.

The MacKenzies Ice Cream Parlors, which we operated in the '60s and '70s, reflected the interest in antiques by the family in art and furnishings, all collected by my wife and myself. Over the years the interest in antiques has changed from operating a small antique shop to collecting such things as art, Hampshire Pottery and Chevy trucks from the late 1950s.

Several years ago the squeaky rubber duck that had adorned the edge of my old bathtub spurred me to find another toy with a squeaker.

This collection, as you will see, shows the wide variety of caricatures and objects that were manufactured with the hope of getting a child's attention. The older rubber toys with a metal squeaker—pristine or in well-loved condition—are most coveted.

The toys shown in this book represent a small part of what is out there for collectors. The squeaky toy your mother saved or the one you kept after the kids grew up may be a prized collectible. I hope this book grabs your attention, just as the squeaker in a rubber toy did when you were a child.

— L.H. MacKenzie

A NOTE ON PRICE RANGE

Values vary immensely according to the condition of the piece, the location of the market, and the overall quality of the design and manufacture. Condition is always of paramount importance in assigning a value. Prices in the Midwest differ from those in the West or East, and those at specialty antique shows will vary from those at general shows. And, of course, being at the right place at the right time can make all the difference.

All these factors make it impossible to create an absolutely accurate price list, but we can offer a guide. The prices reflect what one could realistically expect to pay at retail or auction.

SQUEAKY TOY MANUFACTURERS

The first squeaky toy may have developed from the innocent act of a person stepping on a rubber ball. The air exiting the ball through a small hole probably made an entertaining squeak, hence the first squeaky toy. An early toy made of small bellows and adorned with bird feathers may also be considered the predecessor of squeaky toys.

This edition on squeaky toys contains 301 different squeaky toys, of which 200 have some manufacturer identification and a few where the manufacturer ID was not legible. There are 95 toys with no identification mark to be found on the outside of the toy.

The research of the manufacturers of squeaky toys covered a myriad of sources from the Internet to local and not-so-local libraries, from Chambers of Commerce around the country to countless telephone calls. Some companies were tentative about speaking to me; they were afraid that I was trying to get information for liability problems. Although I was able to pacify some, others were downright rude. One company executive returned my call, not to give me information but rather to gain information on the weather in Vermont; it turns out he has a second home not far from me.

We hope that in subsequent editions of this book, manufacturers will come forward and provide us with more information on the history of squeaky toys.

Manufacturers and Licensees of Squeaky Toys Listed in the Book:

Disney Company, the creator of many cartoon characters, has licensed the manufacturing of its characters to several companies. Today it remains a source of many new toys.

Hasbro a licensee of Disney, is a multi-million dollar company with 13,000 employees and two subsidiaries that produces toys and interactive software. It was founded in 1926 and is located in Pawtucket, Rhode Island. Hasbro stills produces merchandise for Disney.

Shelcore, originally located on Fifth Avenue in New York, first became a licensee of Disney in 1976. They then moved to South Plainfield, New Jersey. Their license to produce Disney products ended in 1988.

Sun Rubber of Barberton, Ohio, was a licensee of Disney from 1941 to 1958. As Sun Products, they again began to make toys for Disney in 1972. In 1981 the company was bought by Wellington Puritan Mills, Inc. of Carrollton, Georgia, which became Wellington Leisure Products in 1983. The last toy Sun Rubber produced for Disney was in 1986.

Ideal Toy Company was owned by CBS, but when Tisch took over from Paley he closed or sold many of the holdings of CBS.

Sanitoy, Inc. was founded in 1917 and is located in Fitchburg, Massachusetts. It merged with Chantex-Pilgrim and currently produces toys under the Nursery Needs label.

Seiberling Latex Products was started in 1921 by Frank Augustus Seiberling, founder of the Goodyear Tire & Rubber Co. Seiberling stepped down as president of Goodyear after the Great Depression forced bank control of the company. Besides developing and manufacturing tires, the company produced other items. It eventually bowed to economic pressure and closed its plant in 1979.

Stahlwood Toy Manufacturing Co. Inc. was founded in 1927 in Youngsville, North Carolina. It closed its doors in the early 1990s.

Names of Companies With No Information:

AIM Plastics (no mention of AIM Plastics, found several under the name AIN:
- AIN (a subsidiary of Thyssen Inc. N.A.) Mt. Vernon, NY
- AIN Plastics of Michigan, Southfield, MI
- Both of these companies are distributors of industrial products, not manufacturers.

Arrow Rubber & Plastic Corp.
- Three names under Arrow, but none of them produced toys:
- Arrow Plastic Products
- Warren, Michigan
- Arrow Plastic Manufacturer
- Elk Grove Village, Illinois
- Arrow Rubber Products (bought out by Gordon Rubber, Derby, CT) Shelton, Connecticut

Ashland Productions Corp.
- Ashland Plastics Inc.
- Ashland, Ohio
- Ashland Products Inc.
- Lowell, Indiana

Eastern Molding Product
- Eastern Molding
- Belleville, New Jersey
- Eastern Molding International, Inc.
- Batavia, New York

Hailand Hall (Tried Holland Hall and Holland Haller)
- Holland Plastics
- Holland, Michigan

Regent Baby Products Corp.
- Regent Sport Corp.
- Hauppauge, New York

Reliable
- Reliable Plastics
- Dunellen, New York
- Reliable Rubber Products Co.
- Bensalem, Pennsylvania
- Reliable Corp. of America Inc.
- Muskegon Heights, Michigan

Reliance
- Reliance Plastics
- Evansville, Illinois
- Reliance Rubber
- Canton, Michigan

Ross Laboratories
- Ross Enterprises
- Newton, Pennsylvania

Star Manufacturing
- Star Manufacturing Inc.
- Little Ferry, New Jersey (never made toys, only custom automotive or electrical parts.)

Uneeda Doll Co., Inc.
- New York, New York

Information Relating to Squeaky Toys not Available at Time of Publication:

Alan Jay @ Clarolyte
Baby Joy
Baby Toy, Co.
Blue Ribbon
Bonnylex (or Bonnytex)
Bye Bye Baby
Cameo KFS
Dank & Squiby
Dreamland Creations
The Edward Mobley Co.
EEGEE
Formulette Co., Inc.
Frandee Corp.
Future Products
Hirsch (or Hersch) Plastic
Irwin Corp.
J.L. Prescott Co.
Jo?? Joy
Jolly Slinker
Kagran
King Features Syndicate
Mark
R
Remple (or Rempel) Enterprises
Rubber King
Rubber Toy
Starwood Toy Manufacturing Co.
Summer (or Sumner) Infant Productions
Terryton
Viceroy
Vinfloat Products

Kangaroo, 8" high. AIM Plastic stamped on toy. Made of vinyl. $40-$50.

Bear Eating Watermelon, 6" high. Alan Jay stamp. Made of vinyl. $20-$30.

Elephant, 6" high. Alan Jay stamped on toy. Made of vinyl. $30-$40.

Tiger with Bottle, 7" high. Alan Jay stamped on toy. Made of vinyl. $30-$40.

Girl with Phone, 7" high. Alan Jay, 1967, stamped on toy. Made of vinyl. $60-$70.

Teddy Bear with Red Cap, 8" high. Arrow Rubber stamped on toy. Made of vinyl. $40-$50.

Bear, 10" high. Arrow Rubber, 1962, stamped on toy. Made of vinyl, with moving head and eyes. $60-$70.

Kangaroo, 9" high. Arrow Rubber stamped on toy. Made of vinyl. $40-$50.

Elephant, 10". Ashland Productions, 1964, stamped on toy. Made of vinyl. $60-$70.

Baseball Player, 5" high. Baby Joy, 1975, stamped on toy. Made of vinyl. $30-$40.

Snoopy© Soap Dish, 2" high. Avon Products/United Features stamped on toy. Made of vinyl. $20-$30.

Black Baby in White Booties, 5" high. Baby Joy, 1979, stamped on toy. Made of vinyl. $60-$70.

Pink Dog, 7" high. Baby Toy, 1967, stamped on toy. Made of vinyl. $30-$40.

Frog, 7" high. Baby Toy, 1968, stamped on toy. Made of vinyl. $60-$70.

Dog, 7" high. Blue Ribbon stamped on toy. Head and legs move. Made of vinyl. $30-$40.

Rabbit, 16" high. Bonnytex, 1961, stamped on toy. Made of vinyl. Head turns. $80-$90.

Baby Graduate Bank, 6" high. Bonnylex Co. stamp on toy. Made of vinyl. $50-75.

Boy in Swimsuit, 8" high. Bye Bye Baby stamp on toy. Made of rubber. $120-$140.

Popeye©, 13" high. Cameo stamped on toy. Made of vinyl. $350-$400.

Kewpie© Doll in Blue, 7" high. Cameo, 1963, stamped on toy. Made of vinyl. $80-$100.

Kewpie© Doll in Red, 7" high. Cameo, 1963, stamped on toy. Made of vinyl. $80-$100.

Kewpie© Doll, 10" high. Cameo stamped on toy. Made of vinyl. $90-$100.

Wizard of O's©, 7" high. Campbell Soup Co. stamped on toy. Made of vinyl. Head moves. $40-$50.

Campbell Soup Kid© (girl), 9" high. Campbell Soup/Ideal Toy Co. stamped on toy. Made of vinyl. $90-$100.

Dog, 3" high. CBS Inc., 1985, stamped on toy. With cloth ears, head turns. Made of vinyl. $20-$30.

Big Bird, 5" high. CBS Inc. stamped on toy. Made of vinyl. $30-$40.

Gabriel©, 5" high. CBS Inc. stamp on toy. Made of vinyl. $10-$20.

Clown, 9" high. Chanlex & Pilgrim stamped on toy. Made of vinyl. $30-$40.

Super Junior (Robin), 6" high. D.C. Comics Inc., 1978, stamped on toy. Made of vinyl. $80-$90.

Clown, 9" high. Chanlex/Pilgrim stamped on toy. Made of vinyl. $30-$40.

Big Boy, 9" high. Dank & Squiby/Big Boy Restaurants stamped on toy. Made of vinyl. $40-$50.

Peter Pan©, 10" high. Disney stamp on toy. Made of vinyl. $50-$60.

Minnie Mouse©, 9" high. Disney stamp on toy. Made of vinyl. $20-$30.

Mickey Mouse, 6" high. Disney stamped on toy. Made of vinyl. $60-$70.

Huey© with Skunk, 4" high. Disney stamped on toy. Made of vinyl. $10-$20.

Mickey Mouse© and Donald Duck© in Inner Tubes, 6" high. Disney stamped on toy. Made of vinyl. $30-$40 each.

Mickey and Minnie Mouse©, 8" high. Disney stamped on toy. Made of vinyl. In original clothes. $300-$400 pair.

Seven Dwarfs in Plastic Carry-all, 5" high. Disney stamped on toy. Made of vinyl. $80-$100 (set).

Dog with red tongue, 5" high. Disney stamped on toy. Made of vinyl. $30-$40.

Mickey Mouse© ABC, 7" high. Disney, 1984, stamped on toy. Made of vinyl. $20-$30.

Thumper from Bambi, 4" high. Disney stamped on toy. Made of vinyl. $10-$20.

Kanga from Winnie the Pooh, 8" high. Disney, 1966, stamped on toy. Made of vinyl. $30-$40.

Huey Duck, 6" high. Disney stamped on toy. Made of vinyl. $20-$30.

Pooh Bear 'Hunny'©, 5" high. Disney stamped on toy. Made of vinyl. $20-$30.

Mickey Mouse© Teething Ring, 5" long. Disney stamped on toy. Made of vinyl. $30-$40.

Baby Duck (one of Donald's nephews), 5" high. Disney stamp on toy. Made of vinyl. $10-$20.

Dopey©, 7" high. Disney stamp on toy. Made of vinyl. $10-$20.

Baby Duck (one of Donald's nephews), 5" high. Disney stamp on toy. Made of vinyl. $10-$20.

Little Mickey©, 6" high, Disney stamp on toy. Made of vinyl. $10-$20.

Pluto, 5" high. Disney stamp on toy. Made of vinyl. $20-$25.

Donald Duck©, 5" high. Disney/Sun Rubber stamped on toy. Made of vinyl. $90-$100.

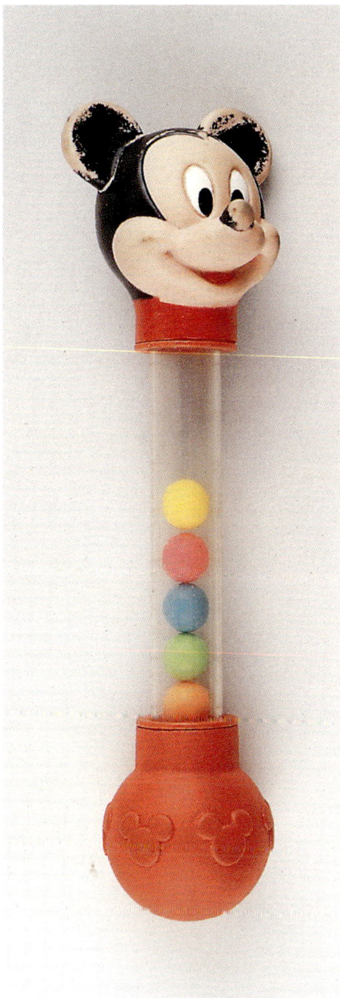

Mickey Mouse© with Rattle and Marbles, 8" high. Disney, 1984, stamped on toy. Made of vinyl. $40-$50.

Minnie Mouse©, 4" high. Disney (China) stamped on toy. Made of vinyl. $10-$20.

Donald Duck©, 6" high. Disney (Dell) stamped on toy. Made of early vinyl. $70-$80.

Minnie Mouse©, 4" high. Disney (Taiwan) stamped on toy. Made of vinyl. $10-$20.

Donald Duck©, 7½ high. Disney/ Dell stamp on toy. Made of rubber. $50-$75.

Baby Duck, 7" high. Disney/Shelcore stamp on toy, 1986. Made of vinyl. $20-$25.

Boy with Bottom Showing, 6" high. Dreamland stamped on toy. Made of vinyl. $60-$70.

Girl with Bottom Showing, 6" high. Dreamland stamped on toy. Made of vinyl. $60-$70.

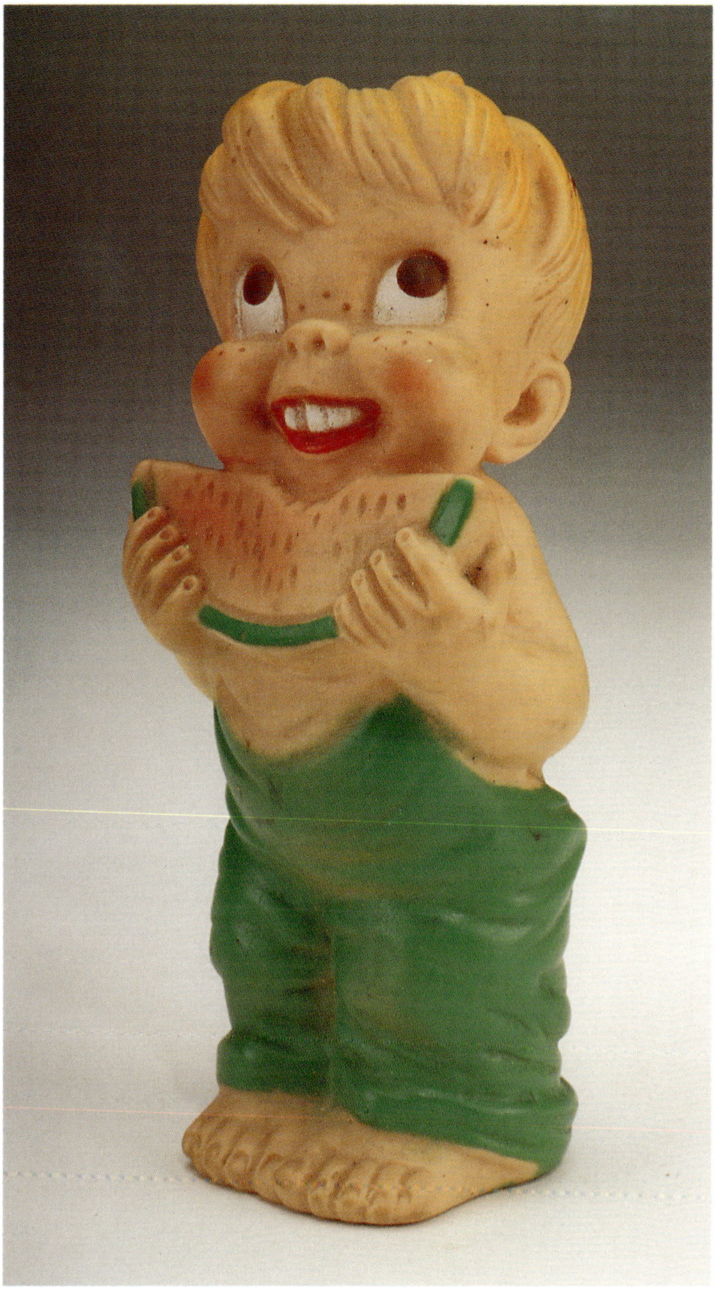

Boy with Watermelon, 7" high. Dreamland stamped on toy. Made of vinyl. $50-$60.

Elephant, 7" high. Dreamland, 1989, stamped on toy. Made of vinyl. $10-20.

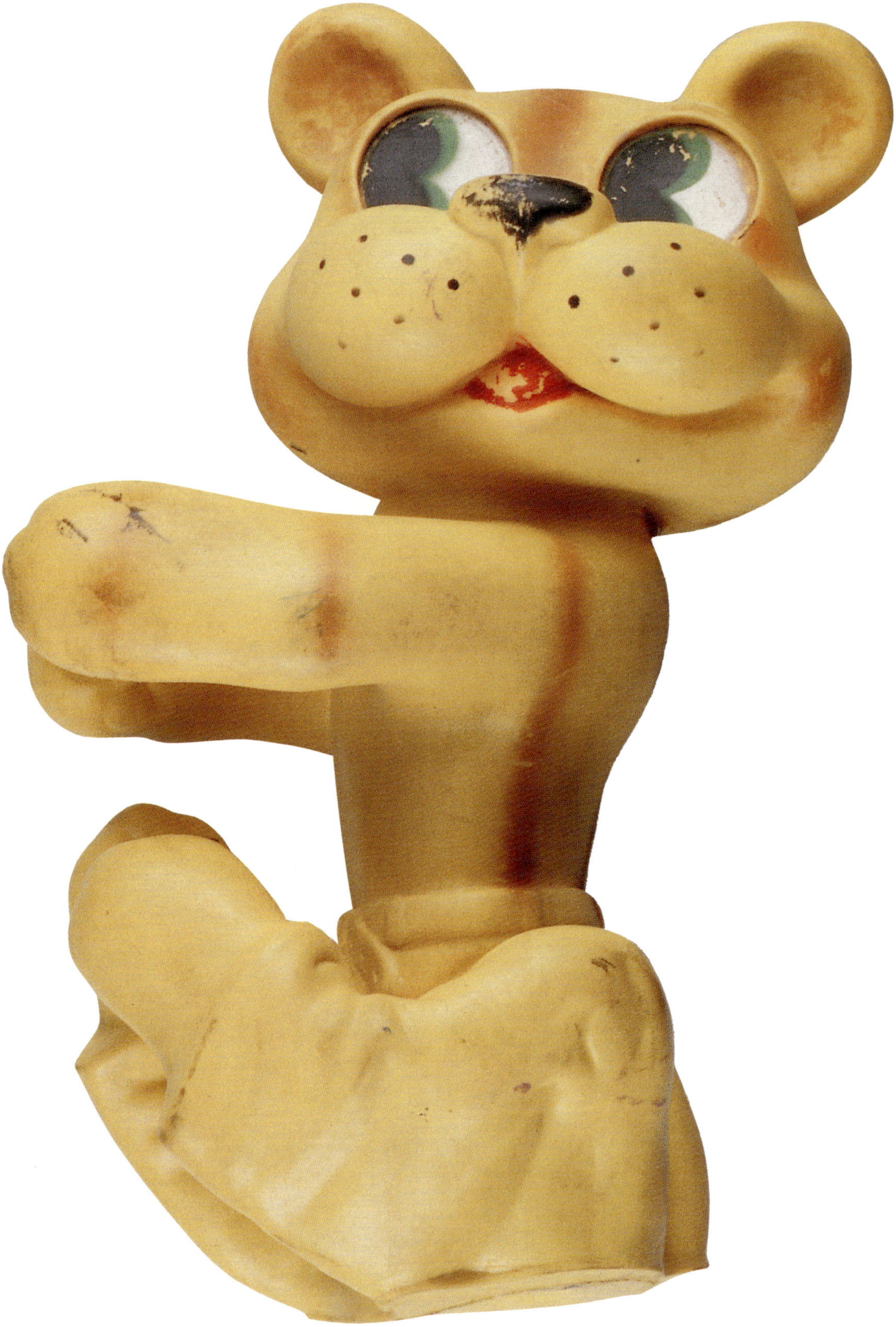

Tiger, 9" high. Baby bottle holder. Dreamland, 1960, stamped on toy. Made of vinyl. $50-$60.

Monkey, 8" high. Dreamland stamped on toy. Made of vinyl. $40-$50.

Dog, 7" high. Dreamland stamped, 1969, on toy. Made of vinyl. $70-$80.

Tomato Man©, 8" high. Dreamland stamp on toy, 1957. Made of vinyl. $75-$85.

Monkey with a Banana, 8" high. Dreamland stamp on toy, 1967. Made of vinyl. $40-$60.

Frog, 3" high. Eastern Molding Product, 1971, stamped on toy. Made of vinyl. $30-$40.

Lamb, 9". Edward Mobley/A.R. Manuf., 1961, stamped on toy. Eyes move. Made of vinyl. $80-$90.

Lamb, 7". Edward Mobley/A.R. Manuf., 1961, stamped on toy. Made of vinyl. $60-$70.

Kitty, 5" high. Edward Mobley/Arrow Rubber stamped on toy. Made of vinyl. $30-$40.

Girl in Yellow, 7". Edward Mobley stamped on toy. Made of vinyl. $70-$80.

"Elephant," 6" high. Edward Mobley/Arrow Rubber stamped on toy. Made of vinyl. $30-$40.

Bean, 10" high. Edward Mobley stamped on toy. Cloth ears and moving eyes. Made of vinyl. $90-$100.

Kitty, 5" high. Edward Mobley stamped on toy. Head turns. Made of vinyl. $50-$60.

Dog, 9" high. Edward Mobley stamped on toy. Head turns. Made of vinyl. $50-$60.

Cat, 9" high. Edward Mobley stamped on toy. Head turns, eyes move, marbles in feet. Made of vinyl. $100-$125.

Boy in Red Suit, 8" high. Edward Mobley/Arrow Rubber stamped on toy. Made of vinyl. $70-$80.

Girl in Pink, 7" high. Edward Mobley/Arrow Rubber stamped on toy. Made of vinyl. $70-$80.

Dog, 6" high. Edward Mobley/Arrow Rubber stamped on toy. Made of vinyl. $40-$50.

Lamb, 6" high. Edward Mobley/Arrow Rubber, 1955, stamped on toy. Made of vinyl. $40-$50.

Girl in Sailor Suit, 12" high. EEGEE stamped on toy. Glass-like eyes, head moves. Made of vinyl. $150-$175.

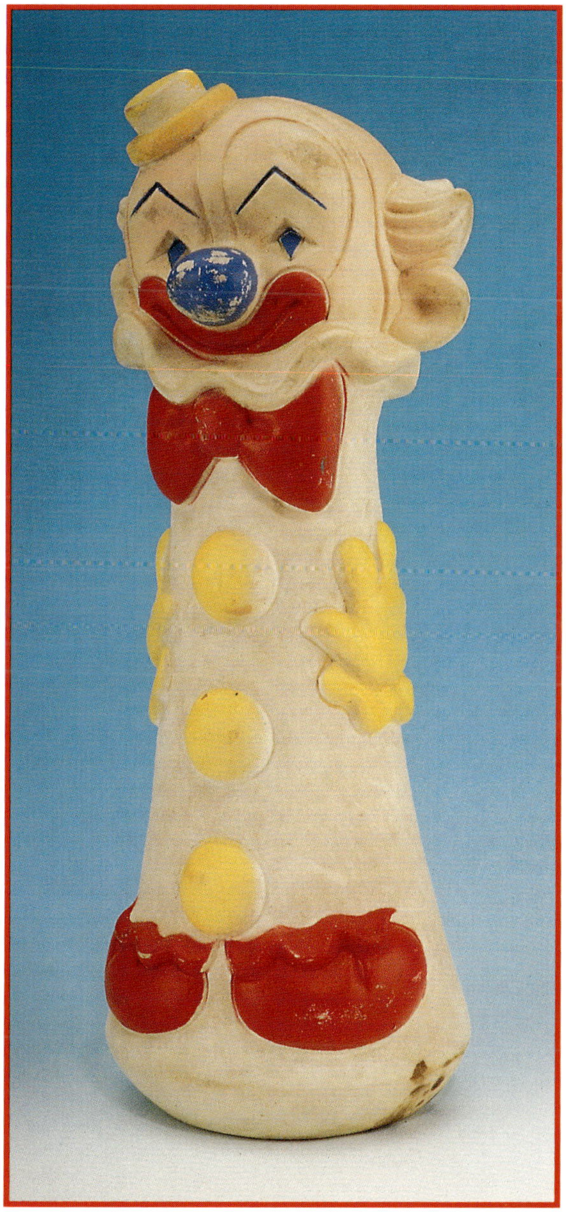

Clown, 8" high. Edward Mobley/Arrow Rubber stamp on toy. Made of vinyl. $40-$50.

Girl, 7" high. EEGEE stamped on toy. Has hair, wets, arms and eyes move. Made of vinyl. $150-$175.

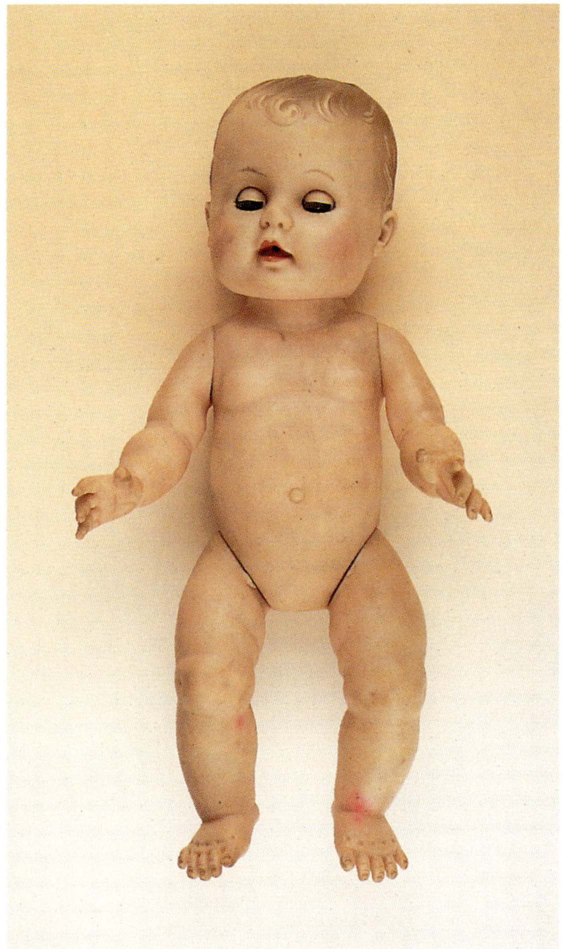

Baby Girl, 12" high. EEGEE 2-55 stamped on toy. Arms, legs, and eyes move; wets. Made of vinyl. $70-$80.

Jolly Jumping Jack, 12" high. Fischer Price Toy Co., 1969, stamped on toy. Made of plastic. $70-$80.

Kangaroo, 9" high. Formulette Co. stamp on toy, 1963. Made of vinyl. $20-$30.

Cowboy, 8" high. Future Products stamp on toy. Made of rubber. $200-$225.

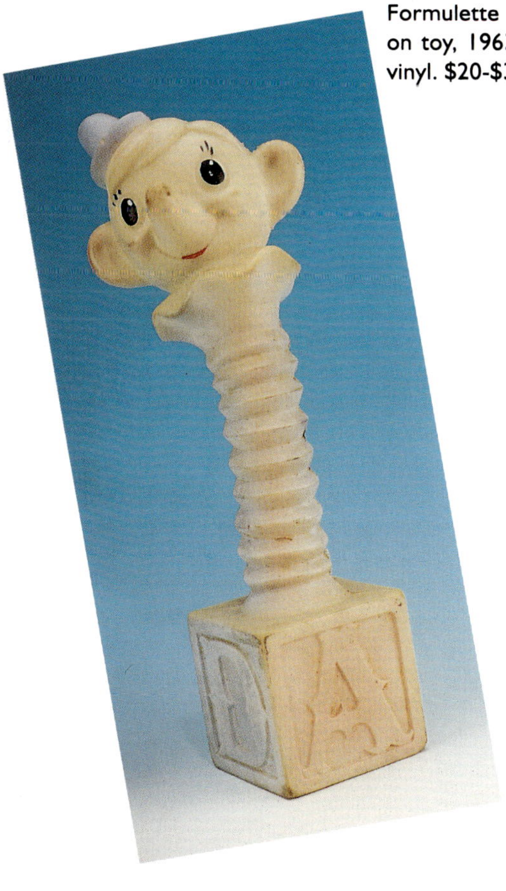

Jack in Box, 10" high. Frandee Corp. stamp on toy. Made of vinyl. $20-$30.

Rabbit, 6" high. Hailand Hall Inc., 1964, stamped on toy. Made of vinyl. $60-$70.

Chipmunk, 4" high. Hasbro/Disney, 1986, stamped on toy. Made of vinyl. $10-$20.

HIPPY©, 5" high. Hirsch Plastic, 1956, stamped on toy. Made of vinyl. In original package. $80-$100.

Rusty, 5" high. Hirsch Plastic, 1956, stamped on toy. Made of vinyl. In original package. $80-$100.

Bunny in Egg Shell, 7" high. Hirsch Plastic stamped on toy. Made of vinyl. $40 $50.

Kitty, 2" high. Holland Haller Inc. stamped on toy. Made of vinyl. $40-$50.

Girl in Sun Glasses, 5" high. Ideal Toy Co., 1966, stamped on toy. Made of vinyl. $80-$90.

Boy in Red Overalls, 5" high. 9 stamped on toy. Made of vinyl. $20-$30.

Baby with Bottle (a Brite-Eye Kid©), 8" high. Irwin stamped on toy. Made of vinyl. $70-$80.

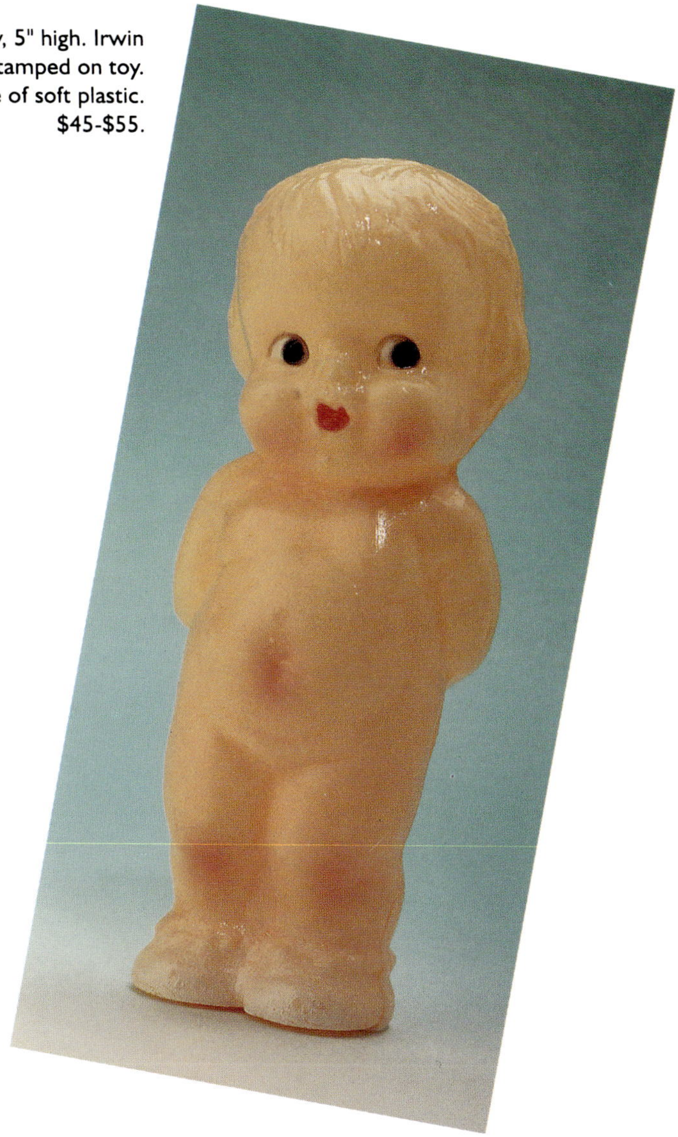

Baby, 5" high. Irwin stamped on toy. Made of soft plastic. $45-$55.

Robin in Original Box, 10" high. Irwin stamped on toy. $150-$175.

Girl, 10" high. Edward Mobley/Arrow Rubber stamped on toy. Glass-like eyes. Made of vinyl. $70-$80.

Dog, 9" high. Irwin stamp on toy. Made of vinyl. $80-$90.

Girl with Raggedy Ann©, 8" high. JL Prescott Co. stamped on toy. Made of vinyl. $30-$40.

Green Cat (Tom from Tom & Jerry?), 6" high. Jo?? Joy stamped on toy. Made of vinyl. $20-$30.

Tiger, 7" high. Jolly Slink PR stamp on toy. Made of vinyl. $10-$20.

Below: Snap, Crackle and Pop©, 7" high. Kellogg Co. stamped on toy. Made of vinyl. $90-$100 each.

Kagran©, 7" high. Kagran stamped on toy. Made of rubber. $200-$225.

Popeye©, 5" high. King Features stamp on toy, 1974. Made of vinyl. $50-$75.

Popeye©, 8" high. King Features Syndicate stamped on toy. Made of rubber. $150-$175.

Bunny, 5" high. M. Dep (made in Italy) stamped on toy. Made of vinyl. $40-$50.

Elephant, 11" high. Mark X stamped on toy. Made of vinyl. Head bobs and eyes roll. $100-$125.

Kewpie© Doll, 8" high. Made in Chile by Dafne stamped on toy. Arms and head move. In original package. Made of vinyl. $40-$50.

Baby Beans, 12" high. Mattel Inc., 1970, stamped on toy. Made of vinyl and cloth. $50-$60.

Clough Pin, 6" high. Made in Taiwan stamped on toy. Made of vinyl. $20-$30.

Beany©, 14" high. Mattel Inc. stamped on toy. Made of vinyl and cloth. $125-$150.

Panda, 11" high. Mattel Inc., 1978, stamped on toy. Made of cloth. $30-$40.

Indian, 8" high. Made in Germany, NY, stamped on toy. $50-$60.

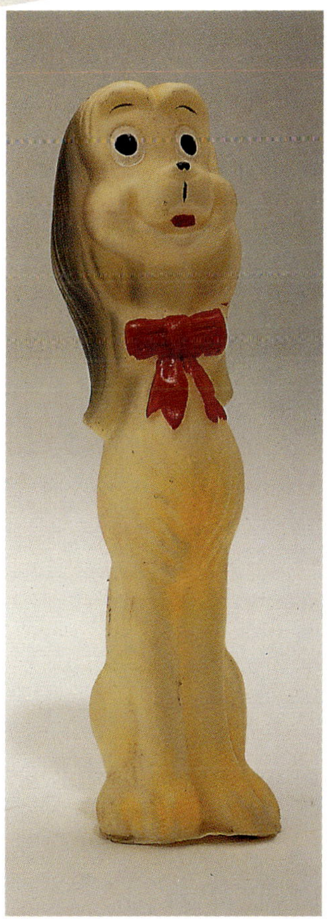

Dog, 6" high. No manufacturer ID. Made in Japan stamped on toy. Made of vinyl. $10-$20.

Baby on Pillow, 6" high. No manufacturer ID stamp on toy. Made of early vinyl. $80-$90.

Big Bird© Look-a-Like on Swing, 5" high. No manufacturer ID stamp on toy. Made of vinyl. $10-$20.

Pig in Red Hat, 6" high. No manufacturer ID stamp on toy. Made of vinyl. $60-$70.

Mouse in Blue Cap, 6" high. No manufacturer ID stamp on toy. Made of vinyl. $15-$20.

Boy with Horn, 6" high. No manufacturer ID stamp on toy. Made of rubber. $50-$60.

Girl, 6" high. No manufacturer ID stamp on toy. Head is made of plastic; body, rattle and squeaky are made of vinyl. $60-$70.

Cat with Green Bow Tie, 5" high. No manufacturer ID stamp on toy. Made of vinyl. $40-$50.

Boy with Bunny, 5" high. No manufacturer ID stamp on toy. Made of vinyl. $10-$20.

Pink Teddy Bear, 8" high. No manufacturer ID stamp on toy. Made of vinyl. $20-$30.

Marky Maypo, 9" high. No manufacturer ID stamp on toy. Made of vinyl. $80-$90.

Lion, 12" high. No manufacturer ID stamp on. Made of vinyl. $60-$70.

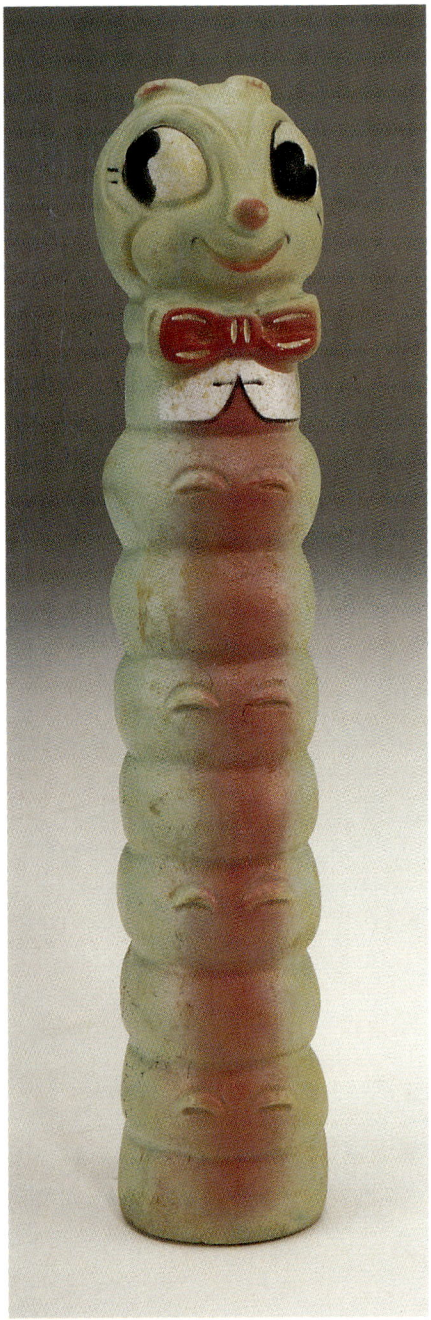

Caterpillar, 9" high. No manufacturer ID stamp on toy. Made of vinyl. $40-$50.

Girl in Blue, 5" high. No manufacturer ID stamp on toy. Made of vinyl. $20-$30.

Panda, 7" high. No manufacturer ID stamp on toy. Made of vinyl. $40-$60.

Hobo, 6" high. No manufacturer ID stamp on toy. Made of early vinyl. $200-$225.

"Howdy Doody©," 12" high. No manufacturer ID stamp on toy. Made of vinyl. $60-$70.

Angel Dog, 5" high. No manufacturer ID stamp on toy. Made of vinyl. $40-$50.

"Hansel," 12" high. No manufacturer ID stamp on toy. Made of vinyl. $150-$175.

Johnny Appleseed, 14" high. No manufacturer ID stamp on toy. Made of vinyl. $250-$300.

Dancing Girl, 10" high. No manufacturer ID stamp on toy. Made of vinyl. $70-$80.

Charlie Chaplin, 8" high. No manufacturer ID stamp on toy. Made of rubber. $800-$1,000.

Green Giant©, 5" high. No manufacturer ID stamp on toy. Made of vinyl. $10-$20.

Snoopy© Look-a-Like, 3" high. No manufacturer ID stamp on toy. Made of vinyl. $10-$20.

Rabbit with Carrot, 5" high. No manufacturer ID stamp on toy. Made of vinyl. $25-$30.

Seal, 3" high. No manufacturer ID stamp on toy. Made of vinyl. $30-$40.

Fish, 4" high. No manufacturer ID stamp on toy. Made of vinyl. $30-$40.

Turtle, 2" high. No manufacturer ID stamp on toy. Made of vinyl. $10-$20.

Car with Face, 3" high. No manufacturer ID stamp on toy. Made of vinyl. $70-$80.

Mouse, 2" high. No manufacturer ID stamp on toy. Made of rubber. $20-$30.

Mouse, 2" high. No manufacturer ID stamp on toy. Made of rubber. $40-$50.

Kewpie© Doll, 5" high. No manufacturer ID stamp on toy. Made of vinyl. $10-$20

Kangaroo, 6" high. No manufacturer ID stamp on toy. Made of vinyl. $10-$20.

Cat, 6" high. No manufacturer ID stamp on toy. Made of vinyl. $20-$30.

Rooster, 6" high. No manufacturer ID stamp on toy. Made of vinyl. $10-$20.

Puppy, 5" high. No manufacturer ID stamp on toy. Made of vinyl. $20-$30.

Kewpie© Doll, 4" high. No manufacturer ID stamp on toy. Made of vinyl. $10-$20.

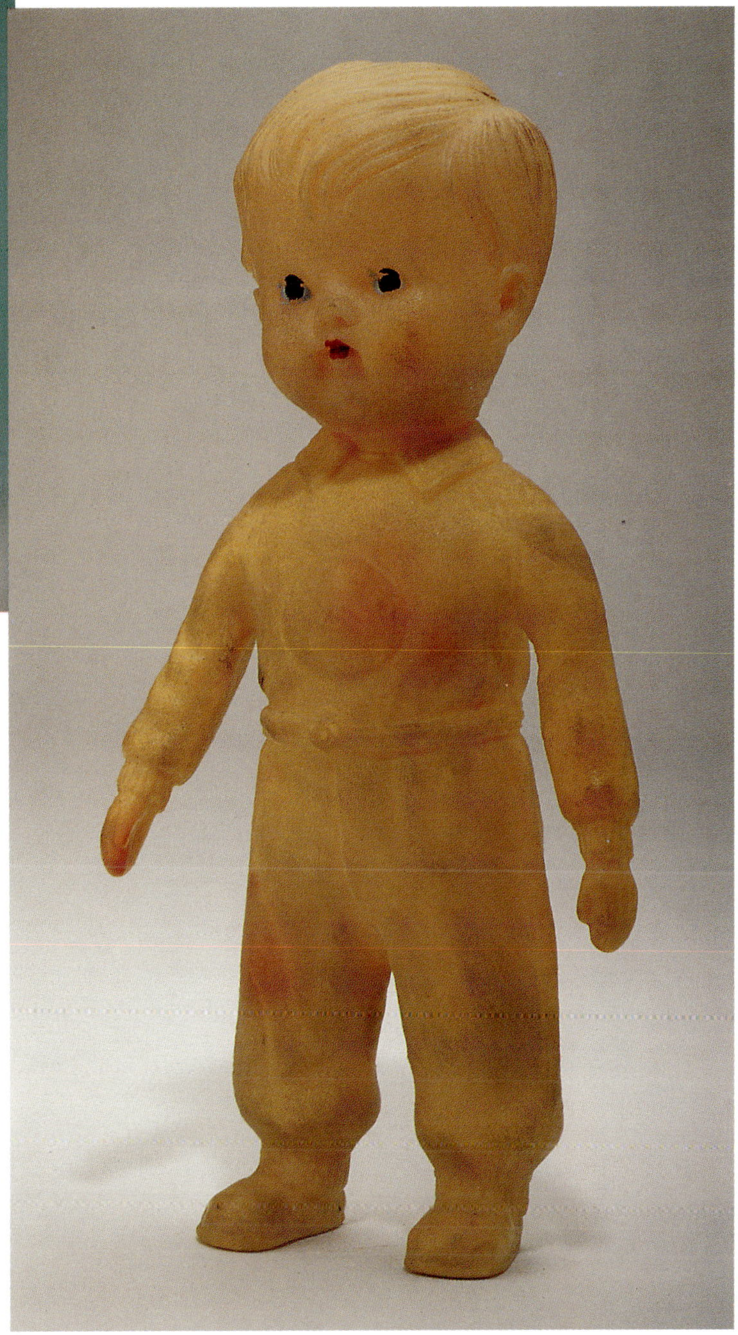

Boy, 9" high. No manufacturer ID stamp on toy. Made of vinyl. $40-$60.

Dog, 5" high. No manufacturer ID stamp on toy. Made of vinyl. $10-$20.

Elephant, 6" high. No manufacturer ID stamp on toy. Made of vinyl. $20-$30.

Baby Girl, 14" high. No manufacturer ID stamp on toy. Cloth pajamas and shoes. Head turns and eyelids close. Made of vinyl. $70-$80.

Elephant, 3" high. No manufacturer ID stamp on toy. Made of vinyl. $10-$20.

Kitty, 6" No manufacturer ID stamp on toy. Made of burlap cloth. $50-$60.

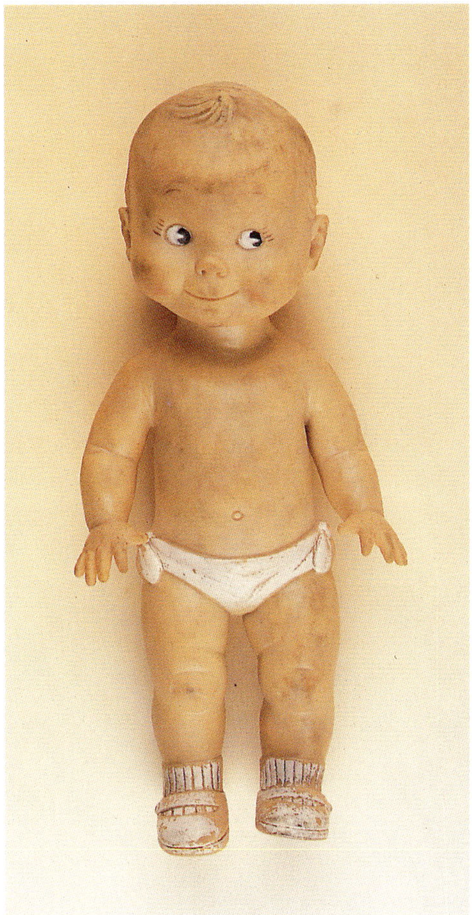

Boy in White Diapers, 9" high. No manufacturer ID stamp on toy. Made of vinyl. $50-$60.

Dog, 6" high. No manufacturer ID stamp on toy. Made of rubber. $80-$90.

Dog, 4" high. No manufacturer ID stamp on toy. Made of vinyl. $10-$20.

Boy, 11" high. No manufacturer ID stamp on toy. Made of vinyl. $70-$80.

Boy (possibly one of Disney's Seven Dwarfs from Snow White?), 11" high. No manufacturer ID stamp on toy. Made of vinyl and cloth. $40-$50.

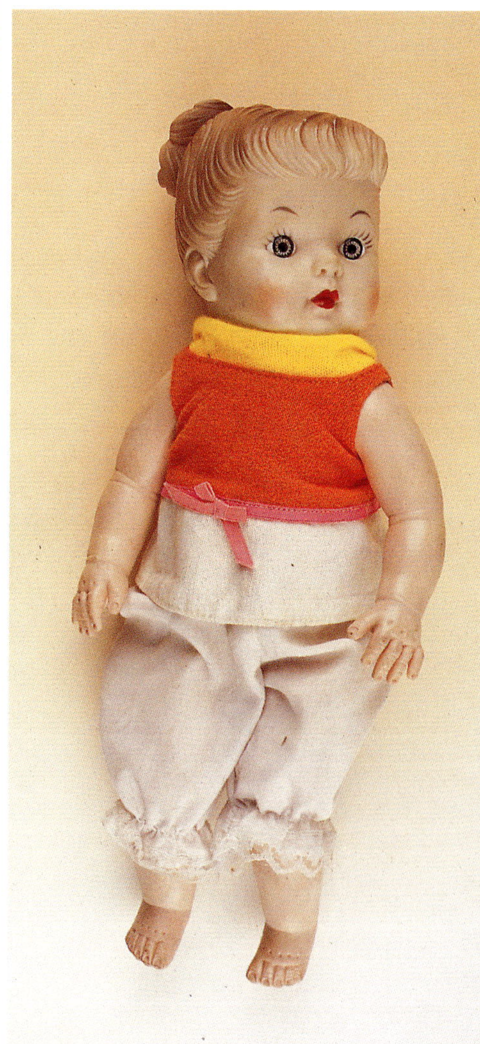

Girl, 12" high. No manufacturer ID stamp on toy. Glass-like eyes. Made of vinyl. $40-$50.

Boy, 11" high. No manufacturer ID stamp on toy. Glass-like eyes. Made of vinyl. $40-$50.

Poodle, 11" high. No manufacturer ID stamp on toy. Made of vinyl. $80-$90.

Bear, 7" high. No manufacturer ID stamp on toy. Made of vinyl. $50-$60.

Santa, 6" high. No manufacturer ID stamp on toy. Made of vinyl. $20-30.

Boy on Swing, 4" high. No manufacturer ID stamp on toy. Made of vinyl. $40-$50.

Prune Man, 4" high. No manufacturer ID stamp on toy. Made of rubber. $50-$60.

Bear, 3" high. No manufacturer ID stamp on toy. Made of vinyl. $10-$20.

Bear, 3" high. No manufacturer ID stamp on toy. Made of vinyl. $10-$20.

Mouse Taking a Bath, 3" high. King Rubber, 1992, stamped on toy. Made of rubber. $10-$20.

Mouse, 3" high. No manufacturer ID stamp on toy. Made of vinyl. $10-$20.

Carrot Man©, 4" high. No manufacturer ID stamp on toy. Made of vinyl. $10-$20.

Girl, 3" high. No manufacturer ID stamp on toy. Made of vinyl. $40-$50.

Bear, 3" high. No manufacturer ID stamp on toy. Made of vinyl. $10-$20.

Pig Bank, 4" high. No manufacturer ID stamp on toy. Made of vinyl. $30-$40.

Pig, 6" high. No manufacturer ID stamp on toy. Made of rubber. $35-$40.

Little Boy, 6" high. No manufacturer ID stamp on toy. Made of vinyl. $10-$15.

Elephant, 5" high. No manufacturer ID stamp on toy. Made of rubber. $60-$70.

Chicken, 4" high. No manufacturer ID stamp on toy. Made of vinyl. $30-$40.

Clown, 7" high. No manufacturer ID stamp on toy. Made of vinyl. $10-$20.

Duck, 3" high. No manufacturer ID stamp on toy. Made of vinyl. $10-$20.

Mouse with Cheese, 7" high. No manufacturer ID stamp on toy. Made of vinyl. $30-$40.

Bunny, 7" high. No manufacturer ID stamp on toy. Made of vinyl. $10-$20.

Bunny, 8" high. No manufacturer ID stamp on toy. Made of vinyl. $60-$70.

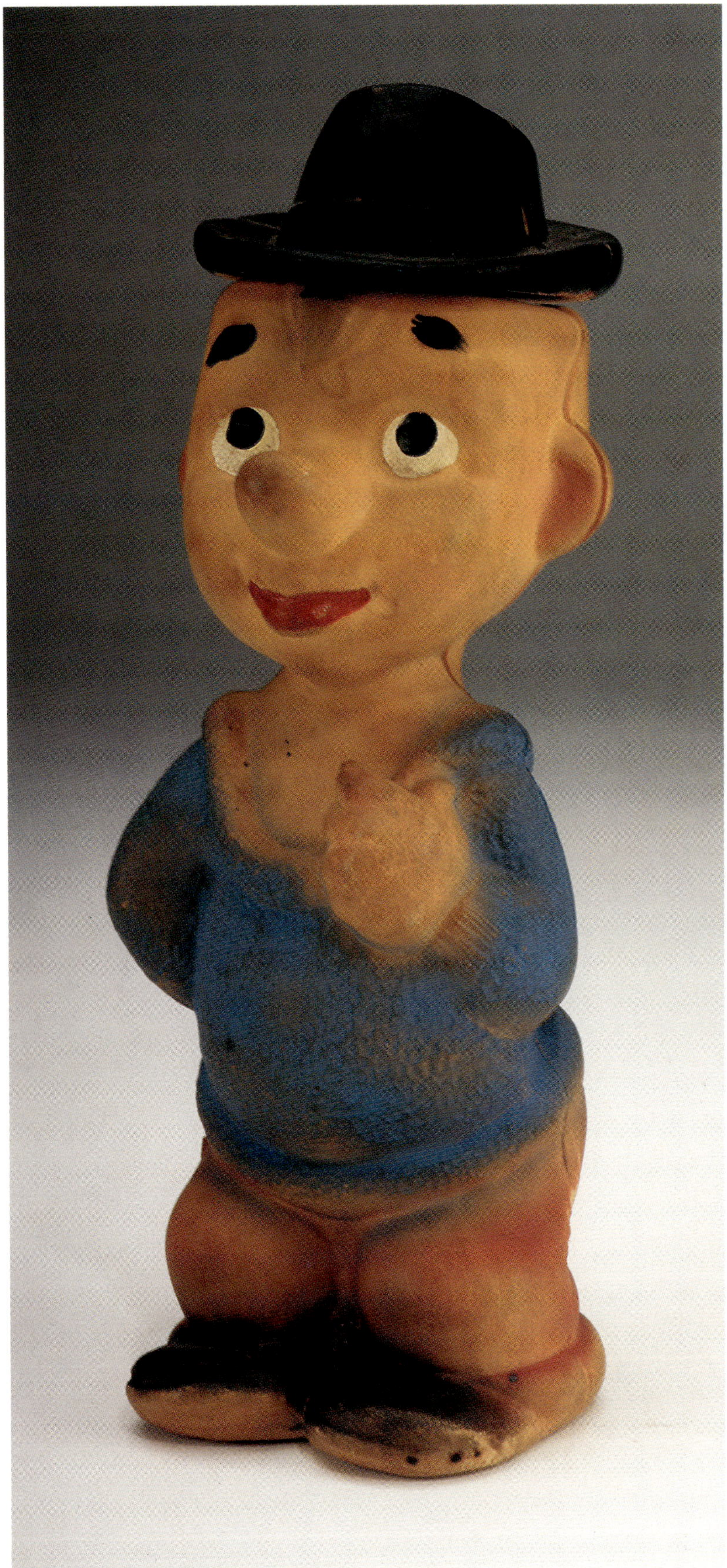

Boy with Black Hat, 9" high. No manufacturer ID stamp on toy. Made of rubber. $150-$175.

Dog in Green Suit, 4" high. No manufacturer ID. 112P80K stamped on toy. Made of vinyl. $10-$20.

Mickey© Look-a-Like, 8" high. No manufacturer ID stamp on toy. Made of vinyl. $25-$30.

Dinosaur, 8" high. No manufacturer ID. Made in China stamped on toy. Made of vinyl. $10-$20.

Bunny, 2". No manufacturer ID. Made in Hong Kong stamp on toy. Made of vinyl. $10-$20.

Snowman, 5" high. No manufacturer ID. Made in China stamped on toy. Made of vinyl. $10-$20.

Kewpie Doll©, 8" high. No manufacturer ID. Made in Hong Kong stamped on toy. Moveable arms and legs. Made of vinyl. $25-$35.

Mac©, Bear with Lollipop, 5" high. No manufacturer ID. Made in Hong Kong stamp on toy. Made of vinyl. $10-$20.

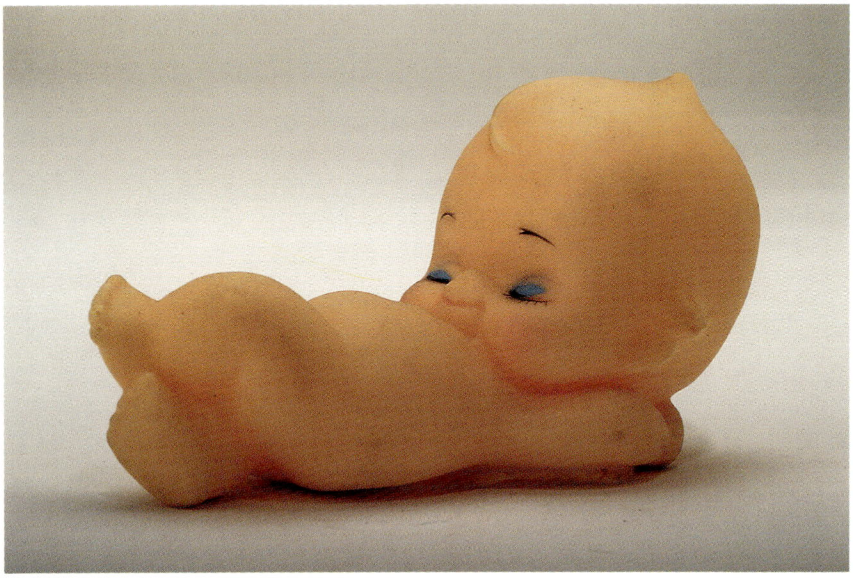

Kewpie© Doll, 3" high. No manufacturer ID. Made in Hong Kong stamped on toy. Made of vinyl. $10-$20.

Bunny with Ball, 6" high. No manufacturer ID. Made in Japan stamped on toy. Made of vinyl. $10-$20.

Stinky the Skunk, 8" high. No manufacturer ID. Made in Japan stamped on toy. Stuffed toy. $50-$60.

Frog, 2" high. No manufacturer ID. Made in Hong Kong stamped on toy. Made of vinyl. $10-$20.

Peasant Girl, 5" high. No manufacturer ID. Made in Japan stamped on toy. Made of vinyl. $10-$20.

Appleman with Rattle, 6" high. No manufacturer ID. Made in Korea stamped on toy. Made of vinyl. $40-$50.

Clown, 8" high. No manufacturer ID. Made in Philippines stamp on toy. Made of vinyl. $10-$20.

Dog, 9" high. No manufacturer ID. Made in Taiwan stamped on toy. Made of vinyl. Head turns and eyes roll. $80-$90.

Raggedy Andy©, 5" high. Made in Taiwan stamped on toy. Made of vinyl. $10-$20.

Monkey, 3" high. No manufacturer ID. Made in Taiwan stamped on toy. Made of vinyl. $10-$20.

Kewpie© Doll, 5" high. No manufacturer ID. Made in Taiwan stamped on toy. Made of vinyl. $10-$20.

Pony, 3" high. No manufacturer ID. Made in Taiwan stamped on toy. Made of vinyl. $10-$20.

Baseball Player, 6" high. No manufacturer ID. Made in Taiwan stamp on toy. Made of vinyl. $10-$20.

Snore Ball for Quiet Sleep, 2" high. No manufacturer ID. Patent pending stamped on toy. Made of rubber. $150-$175.

Santa Dog, 6" high. No manufacturer ID stamp on toy. Made of vinyl. $20-$30.

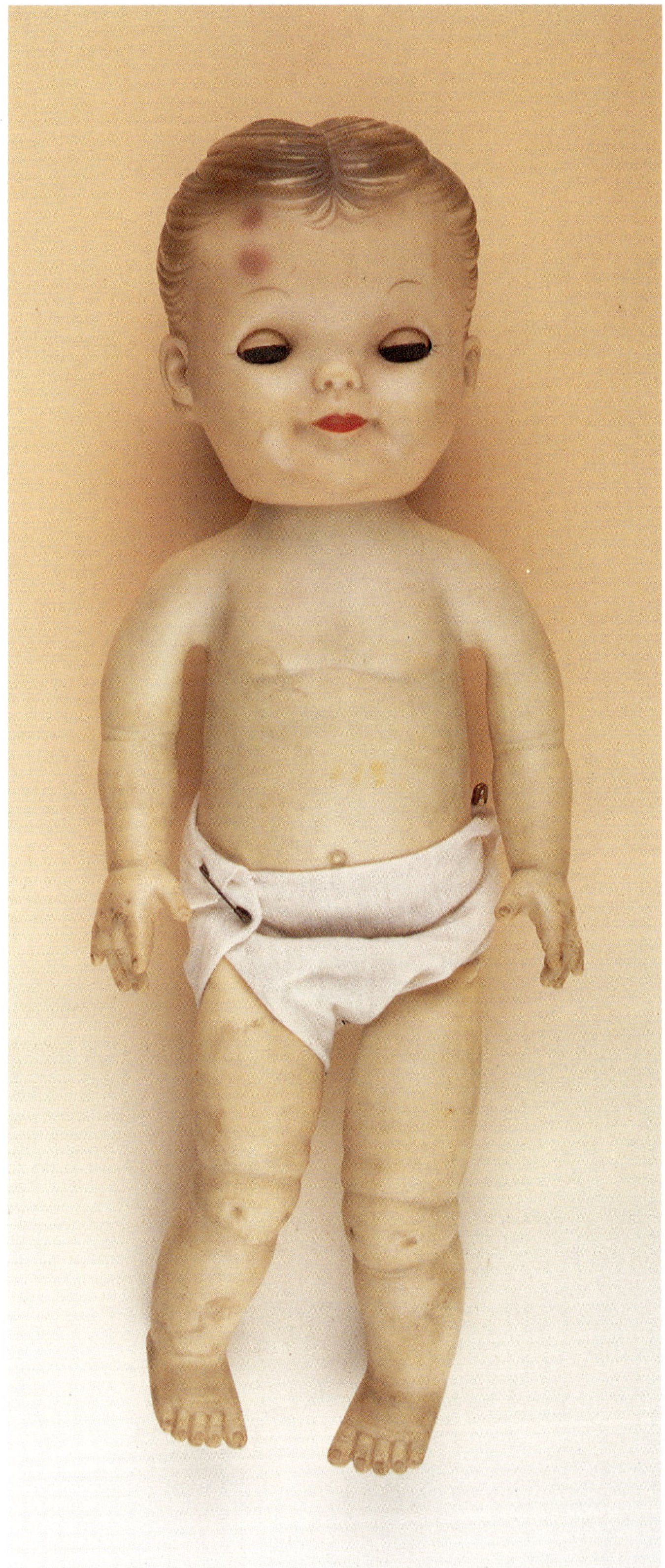

KATIE©, 12" high. Stamped ID not legible. Eyes move. Made of vinyl. $50-$60.

Baby with Teddy, 8" high. R stamped on toy. Made of vinyl. $50-$60.

I am Misty©, 6" high. Rand, McNally & Co., 1960, stamped on toy. Made of vinyl. $90-$100.

Raggedy Ann, 6" high. Regent Baby Products stamped on toy. Made of vinyl. $30-$40.

Dog, 4" high. Reliable stamped on toy. Made of vinyl. $30-$40.

Sheriff, 8" high. Reliance, 1965, stamped on toy. Made of vinyl. $60-$70.

Chipmunk, 6" high. Reliance stamped on toy. Made of vinyl. $30-$40.

Bear with Book, 9" high. Reliance, 1988, stamped on toy. Made of vinyl. $20-$30.

Chick, 6" high. Rempel Enterprises stamp on toy. Made of rubber. $70-$80.

Bunny, 7" high. Rempel Enterprises stamped on toy. Made of rubber. $120-$140.

Duck, 12" high. Rempel Enterprises stamped on toy. Made of rubber. $150-$175.

Indian with Tomahawk, 6" high. Rempel Enterprises stamped on toy. Made of vinyl. $150-$175.

Rooster, 10" high. Rempel Enterprises stamped on toy. Made of rubber. $150-$175.

Cow, 6" high. Rempel Enterprises stamped on toy. Made of rubber. $100-$125.

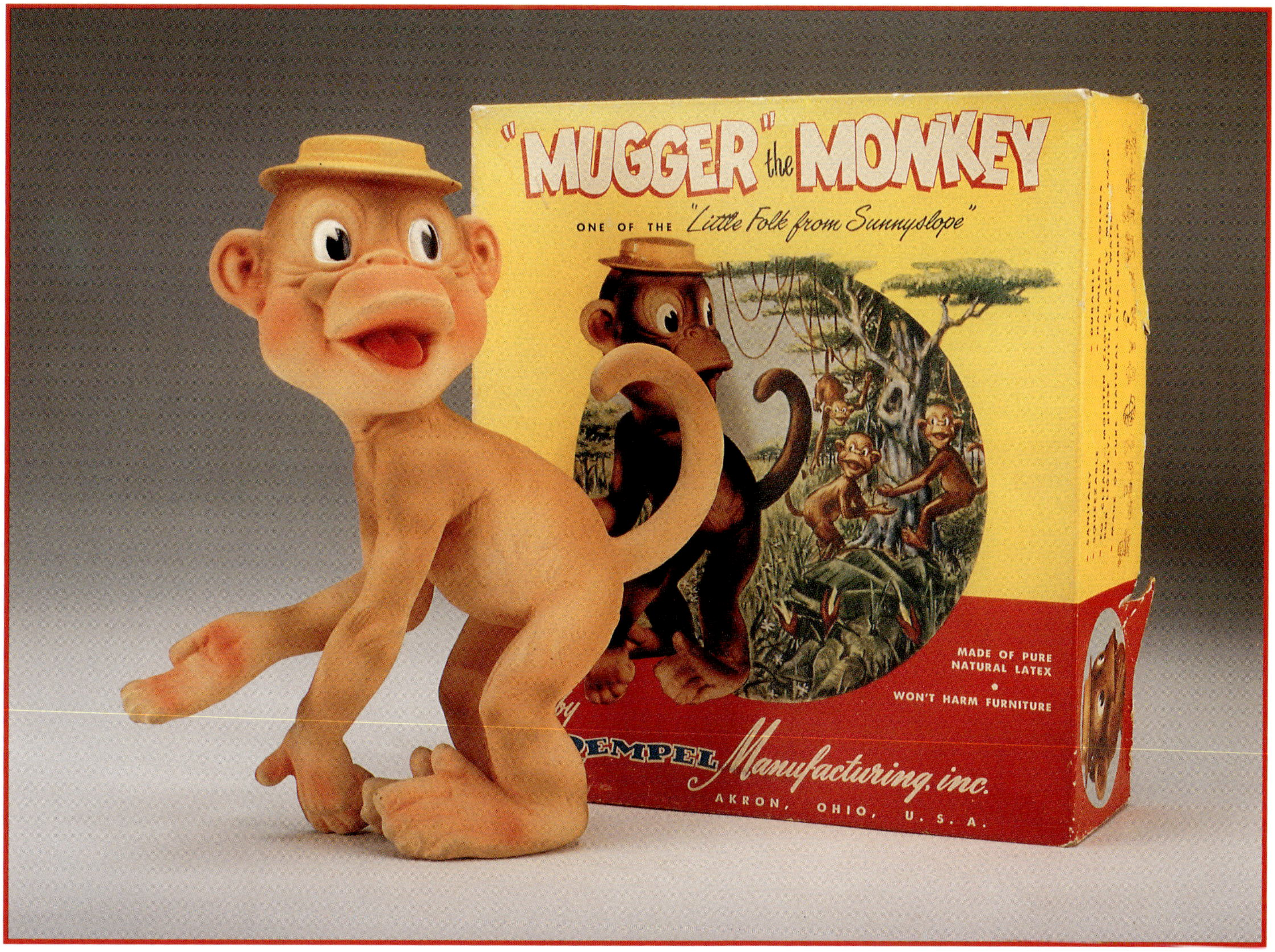

Mugger the Monkey©, 12" high. Rempel Enterprises stamped on toy. Made of rubber. In original box. Mint condition. $500-$600.

Stork, 8" high. Rempel Enterprises stamp on toy, 1957. Made of rubber. $150-$175.

Elephant, 4" high. Rempel Enterprises stamped on toy. Made of rubber. Ears move. $125-$150.

Frog (possibly from *The Wind in the Willows*), 5" high. Rempel Enterprises, 1948, stamped on toy. Made of rubber. $250-$300.

Lamb, 10" high. Rempel Enterprises stamped on toy. Made of rubber. $75-$85.

Lamb, 4" high. Rempel Enterprises stamped on toy. Made of rubber. $75-$85.

Baby, 7" high. Rempel Enterprises stamped on toy. Made of rubber. $100-$125.

Lamb, 7" high. Rempel Enterprises stamped on toy. Made of Rubber. $100-$125.

Dog, 8" high. Rempel Enterprises stamped on toy. Made of rubber. $150-$175.

Kitty in Shoe, 5" inch. Rempel Enterprises stamped on toy. Made of rubber. $70-$80.

Bunny, 12" high. Rempel Enterprises stamped on toy. Made of rubber. $150-$175.

Teddy Bear, 3" high. Ross Lab. stamped on toy. Made of vinyl. $10-$20.

Santa, 11" high. Rempel Mfg. stamp on toy. Made of vinyl. $80-$90.

Yellow Kitty, 4" high. Rubber Toy stamped on toy. Made of vinyl. $20-$30.

Dalmation with Rolling Eyes, 10" high. Saalfield/Sun Rubber, 1950, stamped on toy. Made of vinyl. $100-$125.

Elephant, 7" high. Sanitoy stamped on toy. Made of vinyl. $60-$70.

Boy in Yellow, 8" high. Sanitoy Inc. stamped on toy. Made of vinyl. $20-$30.

Teddy Bear, 4" high. Sanitoy Inc. stamp on toy. Made of vinyl. $50-$60.

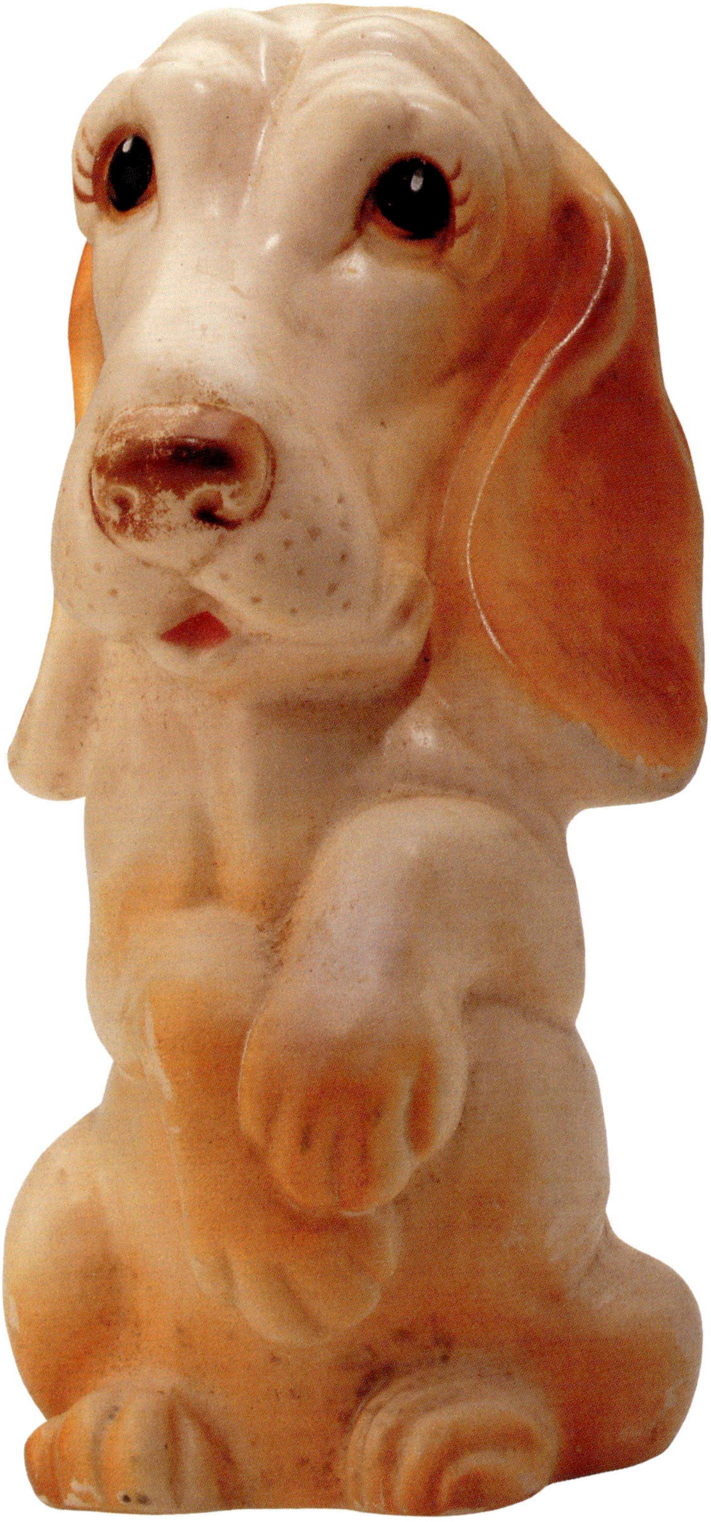

Hound Dog, 8" high. Sanitoy Inc., 1976, stamped on toy. Made of vinyl. $60-$80.

Santa, 8" high. Sanitoy Inc. stamped on toy. Made of vinyl. $40-$50.

Winnie the Pooh with Honey Pot, 3" high. Sears & Roebuck stamped on toy. Made of vinyl. $10-$20.

Bear, 6" high. Seiberling stamp on toy. Made of rubber. $150-$175.

Pelican, 3" high. Shelcore, 1994, stamped on toy. Made of vinyl. $10-$20.

Baldy©, 8" high. Seiberling stamp on toy. Made of rubber. $225-$250.

Lamb, 5" high. Shelcore, 1985, stamped on toy. Made of vinyl. $10-$20.

Dog in Red Hat, 6" high. Seiberling stamped on toy. Made of rubber. $80-$90.

Girl in Bonnet, 7" high. Stahlwood, NY stamp on toy. Made of vinyl. $30-$40.

Dog, 4" high. Stahlwood, NY stamp on toy. Made of vinyl. $30-$40.

Spunky, 8" high. Stahlwood Toy stamped on toy. Made of vinyl. $70-$80.

Santa, 5" high. Stahlwood Toy stamped on toy. Made of vinyl. $30-$40.

Duck, 7" high. Star Manufacturing, 1962, stamped on toy. Made of vinyl. $70-$80.

Santa, 8" high. Starwood Manufacturing stamped on toy. Made of vinyl. $30-$40.

Monkey with Barbell, 4" high. Summer Infant Productions stamped on toy. Made of vinyl and plastic. $20-$30.

Sailor Boy, 8" high. Sun Rubber stamped on toy. Made of vinyl. $80-$90.

Dog, 4" high. Sun Rubber stamped on toy. Made of vinyl. $40-$50.

Kitty in Pink, 6" high. Sun Rubber stamped on toy. Made of vinyl. $25-$30.

Boy in Blue, 8" high. Sun Rubber stamped on toy. Made of vinyl. $70-$80.

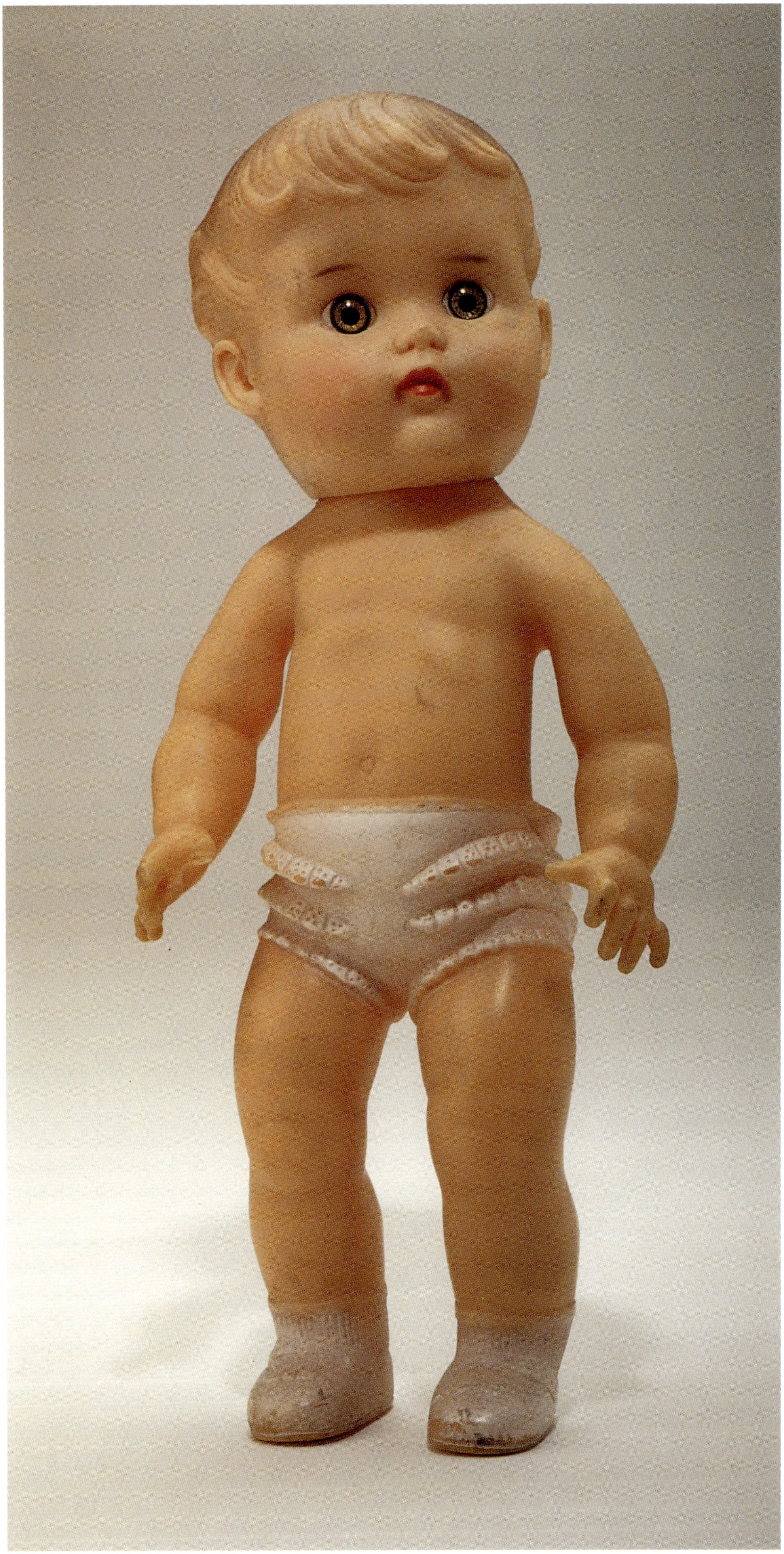

Girl, 10" high. Sun Rubber, 1056, stamped on toy. Made of vinyl. Head moves, realistic eyes. $60-$70.

Opposite page:
Sunbabe SO-WEE©, 9" high. Sun Rubber stamped on toy. Head and eyes move. Made of vinyl. $60-$70.

Chubby Baby, 8" high. Sun Rubber stamped on toy. Made of vinyl. $70-$80.

TOD-L-TIM©, 11" high. Sun Rubber stamped on toy. Made of vinyl. $60-$70.

Lamb, 6" high. Sun Rubber, 1955, stamped on toy. Head moves. Made of vinyl. $50-$60.

Boy, 7" high. Sun Rubber stamped on toy. Made of vinyl. $50-$60.

Girl in Cloth Sun Dress, 8" high. Sun Rubber stamped on toy. Made of vinyl. $50-$60.

Nurse, 5" high. Sun Rubber, 1968, stamped on toy. Made of vinyl. $60-$70.

Elephant, 9" high. Sun Rubber stamped on toy. Head moves. Made of vinyl. $90-$100.

Kitty, 4" high. Sun Rubber, 10D, stamped on toy. Made of vinyl. $30-$40.

Kitty, 4" high. Sun Rubber, 12, stamped on toy. Made of vinyl. $30-$40.

Kitty, 4" high. Sun Rubber, 12D, stamped on toy. Made of vinyl. $30-$40.

Elephant, 10" high. Sun Rubber, 1961, stamped on toy. Made of vinyl. Marbles in feet. $100-$125.

Donkey, 10" high. Sun Rubber, 1967, stamped on toy. Made of vinyl. Marbles in feet. $100-$125.

Oswald Rabbit, 6" high. Sun Rubber (Universal-Walter Lang Creation) stamped on toy. Made of rubber. $100-$125.

Mighty Mouse, 9" high. Terryton stamped on toy. Made of vinyl, with cloth cape. $170-$180.

Bear, 5" high. The First Year stamped on toy. Made of vinyl. $10-$20.

Bear, 3" high. The First Year stamped on toy. Made of vinyl. $10-$20.

Boxer, 6" high. The First Year stamped on toy. Made of vinyl. $10-$20.

Bear, 3" high. The First Year stamped on toy. Made of vinyl. $10-$20.

Fawn, 6" high. The First Year/Kidder Productions stamped on toy. Made of vinyl. $10-$20.

Baseball Catcher, 6" high. The First Year/Kidder Productions stamped on toy. Made of vinyl. $10-$20.

Plumpies™, 6" high. UNEEDA Doll Co., 1968, stamped on toy. Made of vinyl. $70-$80 each.

Opposite page:
Big Girl, 13" high. Uneeda Doll Co. stamp on toy. Made of vinyl. $60-$70.

Opposite page:
Top: Plumpies™, 6" high. UNEEDA Doll Co., 1967, stamped on toy. Made of vinyl. $70-$80 each.

Bottom: Plumpies™, 4" high. UNEEDA Doll Co., 1957, stamped on toy. Made of vinyl. $70-$80 each.

Snoopy© Golfing, 4" high. United Features stamp on toy. Made of vinyl. $10-$20.

Snoopy© with Skis, 4" high, United Features stamp on toy. Made of vinyl. $10-$20.

Snoopy© Sheriff, 5" high. United Features stamp on toy. Made of vinyl. $10-$20.

Snoopy© as Joe Cool, 5" high. United Features stamped on toy. Made of vinyl. $10-$20.

Charlie Brown©, 5" high. United Features stamped on toy. Made of vinyl. $10-$20.

Linus©, 9" high. United Features stamped on toy. Made of vinyl. $70-$80.

Above: Lucy©, 5" high. United Features stamp on toy. Made of vinyl. $20-$30.

Left: Charlie Brown© with Hat, 9" high. United Features stamped on toy. Made of vinyl. $70-$80.

Below: Snoopy© Sleeping, 2" high. United Features stamped on toy. Made of vinyl. $10-$20.

Teddy Bear, 10" high. Viceroy stamped on toy. Moving head, arms, and legs. Made of vinyl. $150-$175.

Andy Panda, 7" high. Vinfloat Products stamped on toy. Made of vinyl. $90-$100.

Road Runner, 8" high. Warner Brothers, 1968, stamped on toy. Made of vinyl and plastic. $250-$300.